Plants in Disguise

Features of Creatures in Flowers and Foliage

Debi Schmid and Lise Hedegaard

2017
Mountain Press Publishing Company
Missoula, Montana

Library of Congress Cataloging-in-Publication Data

Names: Schmid, Debi, 1949- author. | Hedegaard, Lise, 1960- author.
Title: Plants in disguise : features of creatures in flowers and foliage /
 Debi Schmid and Lise Hedegaard 2017.
Description: Missoula, Montana : Mountain Press Publishing Company, 2017.
Identifiers: LCCN 2016056328 | ISBN 9780878426737 (cloth : alk. paper)
Subjects: LCSH: Plant morphology—Juvenile literature. | Mimicry
 (Biology)—Juvenile literature.
Classification: LCC QK49 .S357 2017 | DDC 582—dc23
LC record available at https://lccn.loc.gov/2016056328

PRINTED IN HONG KONG

MP Mountain Press
PUBLISHING COMPANY
P.O. Box 2399 • Missoula, MT 59806 • 406-728-1900
800-234-5308 • info@mtnpress.com
www.mountain-press.com

Illustrator **Debi Schmid** started drawing her kitty Ginger at a young age and has been creating art ever since, earning a degree in art education from the University of Puget Sound. This training was useful during Debi's marketing and communications management career. She and her husband reside in Washington State in a gingerbread house surrounded by fairy gardens, lamb's ears, and cattails.

Author **Lise Hedegaard** is an animal lover and active volunteer at her local Humane Society. With a degree in communications from Pacific Lutheran University, Lise worked as a writer and account executive in the advertising field. She and her husband live in Washington State with their sweet dog and a formerly feral three-legged tabby.

Who Can It Be?

When you wander a woodland forest, skip through a sunlit meadow, or ramble down a dusty path, you might see a furry tail, a bristly beard, or a fuzzy toe. Out of the corner of your eye, did you catch a glimpse of an animal? Or was it a plant in disguise? These wild plants aren't wearing masks or funny noses, but each one displays a feature of a creature. Sometimes you can picture the animal in the leaves, flowers, or seedpods of the plant, but you may need to exercise your imagination to detect the disguise. Many native plants are named after the animals they resemble. We don't know who chose the creative labels, but the people who named the plants were probably Native Americans, early settlers, or pioneering explorers who lived a long time ago.

Once you catch on to the camouflage, you'll be able to recognize these masqueraders wherever you find them. As you thumb through this book, peer into each plant's magnifying glass for a closer look at the evidence. The clues are there for you to discover. You'll learn that some of these undercover plants even have secret powers to heal or help in unexpected ways. Then, next time you're out for a walk and you spy a bill, beard, ear, or foot, take a closer look to see who it can be.

Precisely perched above the bog

Who can I be?

Bird's-Foot *Lotus pinnatus*

A bird's feet help it perform important survival tasks, such as perching on sticks, carrying food, scratching the ground, and even turning eggs. The toes of the bird's-foot plant also have a task: carrying seeds. After the plant has finished flowering, the reddish-brown seedpods develop, looking a little like toes. While most birds have four toes (three facing forward and one pointing back), there can be up to ten "toes" on the bird's-foot plant. Inside each pod are seeds, as in a bean or pea.

See if you can spot the toenails on the seedpod.

Fluffs of fur but a tail that can't twitch

Pretend to pet the tail of this velvety flower spike.

Who can I be?

Cattail *Typha latifolia*

It looks like the tail of a cat, but that brown, furry spike is actually hundreds of tiny flowers. Each flower will produce a teensy seed attached to a fluffy parachute. If you pull a brown spike apart, seeds will billow out as whitish, cottony fluff. Native Americans stuffed the fluff into pillows and mattresses and put it on cuts to help stop bleeding. Cats hate wet feet, so you probably won't find a feline in the wet places where cattails grow.

Horsing around along
a shady stream bank

Who can I be?

Coltsfoot *Petasites frigidus*

This undercover foliage resembles the hoofprint of a high-spirited colt—a young male horse. Coltsfoot leaves are coated on the underside with matted white fuzz and can grow to be 8 to 16 inches across. If a colt's feet were that big, the grown-up horse would be huge. People eat coltsfoot leaf stalks and flower stems. When the leaves are dried and burned, the ash can be used like salt, to make food taste better.

This leathery leaf puts
its best foot forward.

Bill outstretched but not for bugs

Who can I be?

Cranesbill *Geranium bicknellii*

Cranes use their bills to forage for food, but the cranesbill plant has another important use for its long beak. After the delicate flowers are pollinated, they form a rounded fruit with a beak. When the seeds inside are ripe, the beak springs open, snapping the tiny seeds far and wide. If there isn't enough sunlight for the seeds to grow, they can wait in the soil for more than one hundred years. When a wildfire or other disturbance clears away the brush and trees, the seeds will sprout.

**Behold!
This bill is
about to burst.**

The *print is clear.*

Who's been here?

Deer's Foot *Achlys triphylla*

When you bend back the side leaves of this ground-covering plant, the center leaf resembles the hoofprint of a deer. The large leaves rise from underground stems to create a carpet on the floor of moist, shady forests. Native Americans hung bunches of the vanilla-scented leaves in their doorways to keep away flies and mosquitoes. Hikers sometimes rub the fresh leaves on their skin while in the woods during mosquito season.

The center leaflet reveals that a deer's been here.

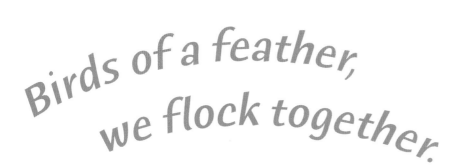

Birds of a feather,
we flock together.

Who can we be?

Ducksbill
Pedicularis ornithorhyncha

With beaks pointing outward like a flock of plucky ducks, the ducksbill plant grows in mountain meadows watered by melting snow. While ducks use their flat bills to fish for food, these petite pink or purple bird heads do their best to attract bumblebees. The bees transport pollen from flower to flower, an essential step in producing seeds. Without bees, the ducksbill would be gone in one quick quack.

Observe the curve of the duck's long neck.

A protruding proboscis
but no pointy tusks

Who can I be?

Ears to you!
Peek at the
prominent
petals.

Elephant's Head   *Pedicularis groenlandica*

The flower spike of the elephant's head plant has space for a whole herd of pink pachyderms! (*Pachyderm*, which means "thick skin" in Greek, is another name for the elephant.) The upper lip of the bloom looks like the head and trunk of an elephant, while the side petals flop like his ears. Imagine the noise in the meadow if all these trunks trumpeted in time! Proboscis (pronounced pro-BA-*skiss*) is another word for the elephant's strong snout.

Racks of antlers
atop straight stems

Who can I be?

The hunt is on for the floral horns of an elk.

Elkhorn

Clarkia pulchella

Whoever named this wildflower probably didn't know that those pointy prongs on top of a male elk's head are called antlers, not horns. Antlers are shed in late winter, but horns are permanent, continuing to grow throughout an animal's life. Male elk grow new antlers each spring. The elkhorn plant dies every winter, but its seeds sprout in spring and produce new antler flowers.

The tuft of my tail tells
you where I am.

Who can I be?

Focus on the
flowers if
you want
to find the
fox.

Foxtail *Alopecurus aequalis*

When this grass sprouts its bushy summer flowers, it looks like a family of foxes strolling along the stream. The seeds have a bristle, called an awn, that can stick to the fur of passing animals. These hitchhiking seeds then find new places to grow. Be sure to remove the bristles from your dog's fur and from your socks and shoes. You don't want a foxtail roaming your yard!

Wispy whiskers that shouldn't be shaved

Who can I be?

Put hundreds of these tiny male flowers together to create the weird beard of a goat.

Goat's Beard *Aruncus dioicus*

Both male and female goats have beards, but a single goat's beard plant has either male flowers or female flowers. Both types of tiny flowers cluster on long stems, forming creamy-white wisps that look like spaghetti to some people and a goat's beard to others. For many Native American tribes, the goat's beard plant was as good as a pharmacy. They used the roots to make medicines to treat colds and a variety of diseases.

Earthbound but with the beard of a bird

Who can I be?

Hawksbeard

Crepis elegans

After this plant's sunny yellow flowers have faded, the pale tufts resemble the bristly feathers under the chin of a hawk. Though the hawksbeard's taproot keeps it from soaring like a bird, each plant can send 50,000 seeds flying through the air. (A taproot is a long, skinny root that grows deep into the soil.) Like its relative the dandelion, the hawksbeard spreads easily and is sometimes considered a weed.

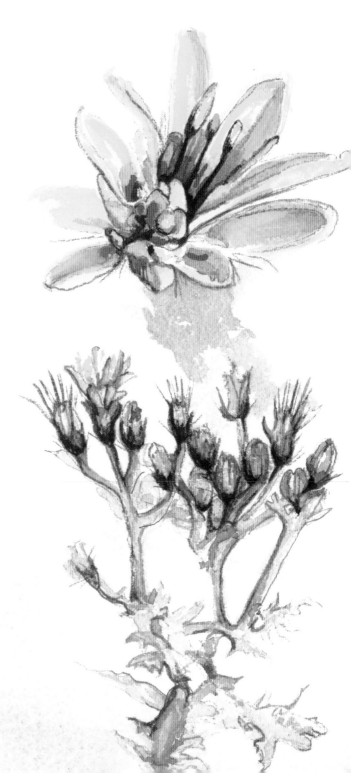

The flowers fade to
the bits of a beard.

Maybe a swish
but not a whinny

Who can I be?

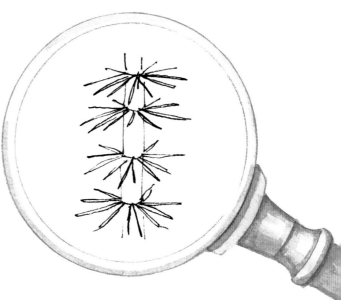

Study the stem
and branches as
you trail the tail.

Horsetail
Equisetum species

Horsetail's thin branches twirl out from the center stem like fireworks. When the branches are fully grown, the straight stalk looks like a horse's tail. The hollow stems are coated with silica, a rough and scratchy substance that makes the plant useful for scrubbing metal pots and tin cups. When the dinosaurs were alive, horsetails were the size of trees!

Hound's Tongue   *Cynoglossum grande*

The fuzzy leaf of this plant is shaped like a dog's tongue, but it will never lick you on the ankle as you walk in the woods. People used to think you could keep a dog from attacking by putting one of these leaves in your shoe. We don't know if this worked, but some varieties of hound's tongue smell bad, so that may have kept the dogs away! The seeds are covered with hooked bristles that can stick to clothes and animals, spreading the plant to new locations.

Did you see that? This plant just stuck out its tongue!

Frisky felines, furred
and first to appear

Who can we be?

Find the fuzz at the flower's center.

Kittentails *Synthyris missurica*

This kindle of kittens, with their tails in the air, is one of the mountain forest's earliest bloomers. (*Kindle* is another word for a litter of young cats or other baby animals.) Each kittentails plant has several flowering stalks, some curving like the tail of a contented cat. The flower parts that tickle the bright blue petals give each cluster of flowers a fuzzy look.

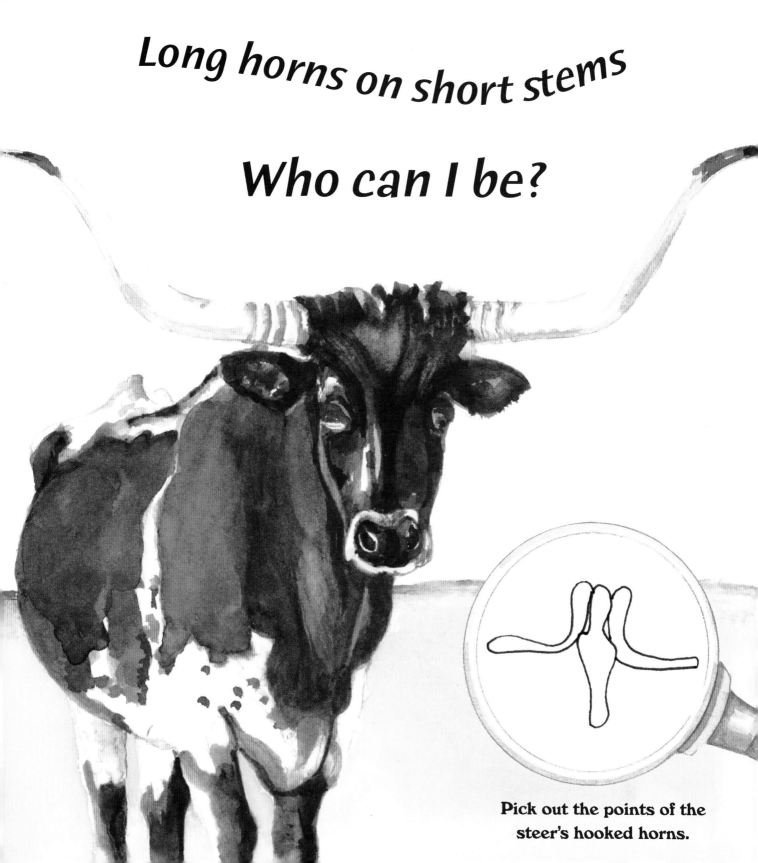

Long horns on short stems

Who can I be?

Pick out the points of the steer's hooked horns.

Longhorn Steer's-Head

Dicentra uniflora

These delicate, pinkish blossoms resemble a steer's head, with horns curved up and inward. Home on the range for this plant is a rocky slope near the edge of a melting snowbank. Look closely, because this herd is not much taller than an all-beef burger. While the flower is just half an inch wide, the horns of longhorn cattle are up to seven feet wide, tip to tip. Longhorns are descendants of the first cows in the Americas, brought in by Christopher Columbus.

Bright button eyes in a
funny little face

Who can I be?

Monkey Flower

Mimulus guttatus

Many flowers seem to have faces. If you squint, the blossoms of monkey flower mimic the face of a monkey, with two big eyes, round cheeks, and a heart-shaped forehead. Because this plant absorbs salt from the soil, early settlers used the leaves and stems as a salt substitute to flavor meat. The monkey flower has also been very useful to scientists who study plant evolution.

Looks like this flower face is up to some kind of monkey business.

Tread on my tail
and I'll scarcely squeak.

Who can I be?

Mousetail

Myosurus apetalus

Mice rely on their tails for balance and body temperature control, but the mousetail plant has a different tale to tell. This marsh-dwelling plant was named for its seed head, because the stacked seed compartments resemble the scaly texture of some rodents' tails. The slender leaves are also the size and shape of a mouse's tail. The mousetail plant has been used by some Navajo people to protect against witches.

Eek! This flower spike might be the tail of a mouse!

Poised to perk up

Who can I be?

Mule-Ears

Wyethia amplexicaulis

This plant is all ears, but it can't hear a word. The long, green leaves, shaped like the ears of a mule, are food for mule deer in the spring when the foliage is soft. Later in the year, elk and deer munch the sunflower-like blooms. Native Americans cooked the sturdy roots of the mule-ears plant in sealed pits heated with hot stones to make a sweet treat. Mule-ears often cover an entire dry field or hillside, providing safe places for birds and small mammals to live.

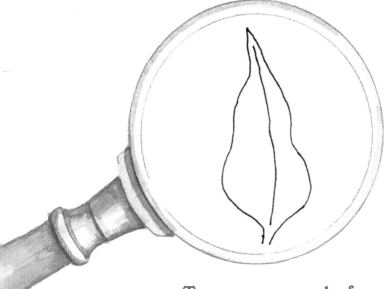

Turn over a new leaf, and there's another ear.

Three tiny toes, tracks on the terrain

Who can I be?

Partridgefoot

Luetkea pectinata

The divided leaves of this summer bloomer resemble the three-toed footprint of a partridge. Partridge tracks are common because these birds walk more than they fly. Partridges never nest in trees, so they would probably feel at home in the partridgefoot's habitat—high in the mountains where trees don't grow. Partridgefoot leaves are crowded together at the base of slender flower stems, creating a bright green mat that covers rocks and soil after the snow melts.

Glance at the stance of the partridgefoot leaves.

I probably wouldn't purr
if you tickled my toes.

Who can I be?

Pussytoes *Antennaria rosea*

The small, fuzzy flower heads of the pussytoes plant resemble a cat's paw, but without the claws. White or pink, plush and pretty, these flowers attract butterflies, but things with wings would surely stay away if those toes were attached to a cat! Look at the bunched blossoms from above, and then examine the underside of your cat's paws.

Inspect the plush pad of this pussytoes paw.

Soft and silky, heads or tails

Who can I be?

Squirreltail *Hordeum jubatum*

These feathery flower heads, as fine and fluffy as a squirrel's tail, are not just for show: each plant can produce 200 seeds. When the wind wafts the seeds onto areas damaged by fire, water, or the actions of humans, the seeds sprout into new plants. Squirrels have their own way of helping plants grow. The bushy-tailed rodents bury seeds and nuts to save for later but sometimes forget to dig the food up. The uneaten seeds sprout and become seedlings in the spring.

The seeds are squirreled away inside this flower plume.

You Can Be Wise Beyond Disguise

Plants are considered *native* when they have existed in an area for many years without humans bringing them there. Some have adapted to very specific habitats, with unusual weather patterns or soil conditions. Some native plants live only in a small area; others are spread across the United States and around the world.

We call native plants *wildflowers* if they are showy and live in the wild without help from humans. A wildflower can be considred a *weed* if it has spread to an area where it isn't wanted. Weeds often multiply quickly and crowd out native plants.

Different plants sometimes have the same common name, so scientists have given each plant a one-of-a-kind scientific name. For example, the common name *hound's tongue* is used for many related plants, including a weed poisonous to horses and cows (*Cynoglossum officinale*) and a cheerful blue wildflower native to the Pacific Coast (*Cynoglossum grande*). The first part of their scientific name is the same (*Cynoglossum*), revealing that they are closely related to one another. The poisonous hound's tongue is native to Europe, where it is considered a wildflower. After it was accidently brought to North America, this problem plant spread quickly, becoming a weed.

Even plants that are related may grow in very different areas and have different characteristics. The plants in this book are common in many northern states and on the West Coast. Though some of them may not grow in your area, you might find a close relative!

In this book, you've read about how some people have used these plants for food or medicine, but don't try this at home. When you see a native plant in the wild, it's usually better to look instead of touch. Another living thing may be depending on that plant or flower.

Where Can I Be?

The illustrations in this book depict individual species that grow in specific parts of the United States. Here are the regions where you could find each of these plants in disguise.

BIRD'S-FOOT (*Lotus pinnatus*)
Perching on the West Coast and in Idaho at low elevations

CATTAIL (*Typha latifolia*)
Stalking marshes and ponds throughout the continental United States

COLTSFOOT (*Petasites frigidus*)
Trotting down the West Coast from Alaska to California and all the way across the northern United States

CRANESBILL (*Geranium bicknellii*)
Flinging seeds throughout much of North America, except along the southern border and lower Great Plains

DEER'S FOOT (*Achlys triphylla*)
Carpeting moist areas from British Columbia south to California

DUCKSBILL (*Pedicularis ornithorhyncha*)
Flocking from Alaska to Washington in high mountain meadows

ELEPHANT'S HEAD (*Pedicularis groenlandica*)
Stampeding through the western half of the United States

ELKHORN (*Clarkia pulchella*)
Charging around open slopes from British Columbia to Oregon and east to South Dakota

FOXTAIL (*Alopecurus aequalis*)
Roaming wet places in most of the continental United States, except the southeast region

GOAT'S BEARD (*Aruncus dioicus*)
Sprouting whiskers from Alaska south to California, and in the east

HAWKSBEARD (*Crepis elegans*)
Seeds on the wing in Alaska, British Columbia, Montana, and Wyoming

HORSETAIL (*Equisetum* species)
Swishing on the Pacific Coast and east to Idaho

HOUND'S TONGUE (*Cynoglossum grande*)
Romping in the woods from southern British Columbia to southern California

KITTENTAILS (*Synthyris missurica*)
Waving in the shade on the West Coast and in Idaho and Montana

LONGHORN STEER'S-HEAD (*Dicentra uniflora*)
Grazing in high places along the West Coast and east to Montana and Colorado

MONKEY FLOWER (*Mimulus guttatus*)
Monkeying around from Alaska to California, east to the Dakotas, and in Colorado and New Mexico

MOUSETAIL (*Myosurus apetalus*)
Squeaking in from British Columbia to California and east to Wyoming, plus the Dakotas

MULE-EARS (*Wyethia amplexicaulis*)
Perking up from Washington south to Nevada and east to the Rocky Mountains

PARTRIDGEFOOT (*Luetkea pectinata*)
Trekking through mountainous areas up and down the West Coast and in the Rockies

PUSSYTOES (*Antennaria rosea*)
Pussyfooting high and low along the Pacific Coast, east to Quebec and New Mexico, and in North Dakota and Nebraska

SQUIRRELTAIL (*Hordeum jubatum*)
Scampering in full sun throughout the continental United States, except in the extreme southeast

Explore More!

If you would like to learn more about native plants, here are a few resources we found helpful.

Common Edible and Useful Plants of the West, by Muriel Sweet
   Easy-to-read details about the edible and medicinal uses of trees, shrubs, vines, herbs, and water plants of the western United States.

Burke Museum Herbarium at the University of Washington
   http://biology.burke.washington.edu/herbarium/imagecollection.php
   This website has a large database of plant images and location information.

Handbook of Nature Study
   http://handbookofnaturestudy.com/join-us-today/
   Fun and educational activities for children.

Lady Bird Johnson Wildflower Center
   http://www.wildflower.org/explore/
   Home of the Native Plant Information Network (NPIN), which has a goal to assemble and disseminate information that will encourage the sustainable use and conservation of native wildflowers, plants, and landscapes throughout North America.

The United States Department of Agriculture Celebrating Wildflowers
   http://www.fs.fed.us/wildflowers
   Go to Special Features, and then to Plant of the Week to look up specific plants. Along the way, enjoy fun coloring sheets and word searches for kids, and excellent resources for parents and teachers.

USDA Plants Database
   http://plants.usda.gov/java/
   A comprehensive database of plants across the US.